No caterpillars or butterflies
were harmed in creating this
Adventure Book!
Maddison and Mimi continue to learn
all that they can about the life cycle
of the butterfly. They learn about
blooming plants perfect for attracting
butterflies, and which host plants
are perfect for each species of
MUNCHING caterpillars
to grow big and fat!
For more information check out
our website at:

www.thecaterpillarsthatgrew.com

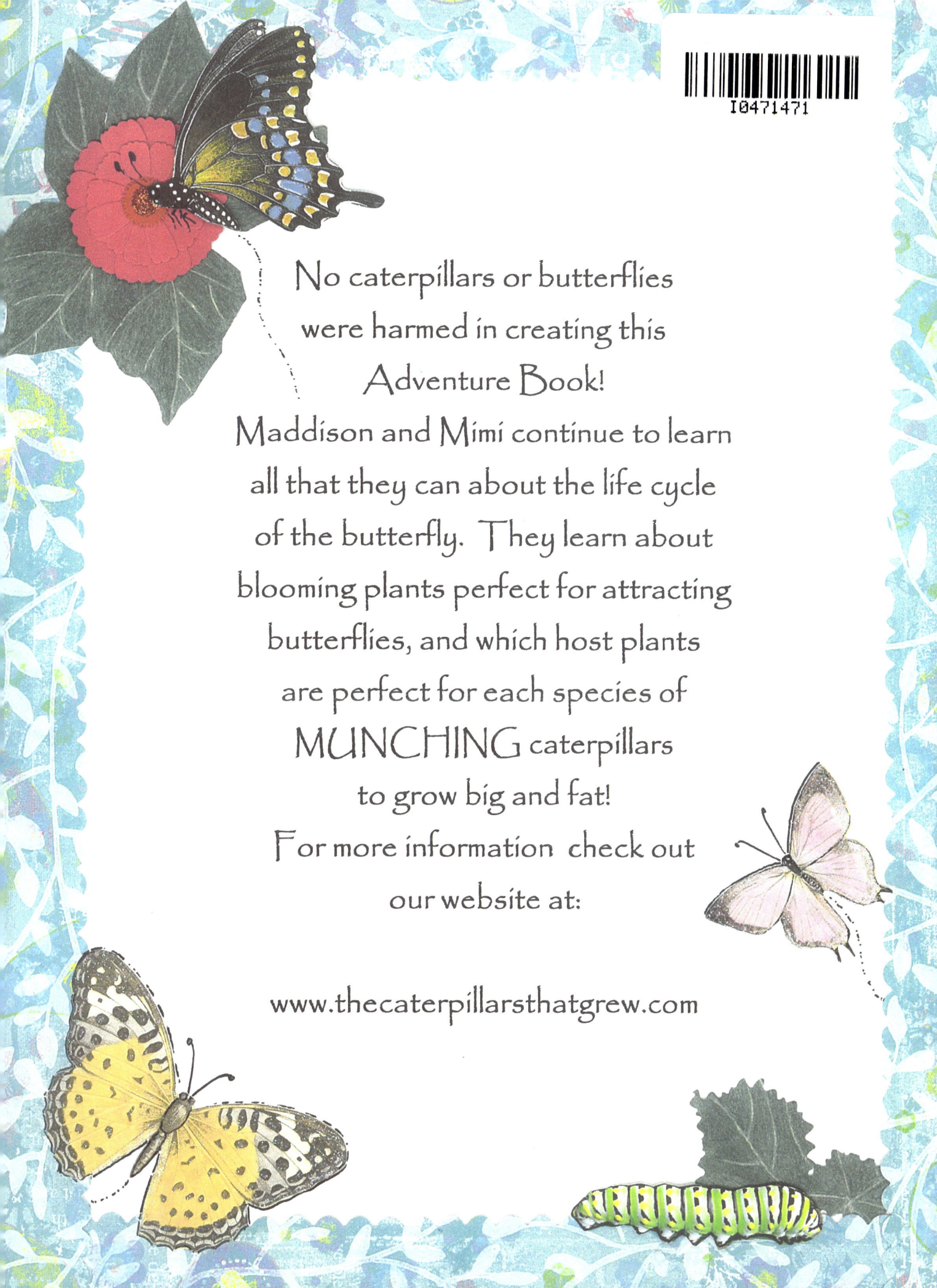

This Book Belongs To:
_____

Order this book online at www.trafford.com
or email orders@trafford.com

Most Trafford titles are also available at major online book retailers.

© Copyright 2010 Robin Murray.

All rights reserved. No part of this publication may be reproduced, stored in a retrieval system, or transmitted, in any form or by any means, electronic, mechanical, photocopying, recording, or otherwise, without the written prior permission of the author.

Printed in Victoria, BC, Canada.

ISBN: 978-1-4269-2240-4

Our mission is to efficiently provide the world's finest, most comprehensive book publishing service, enabling every author to experience success. To find out how to publish your book, your way, and have it available worldwide, visit us online at www.trafford.com

Trafford rev. 3/16/2010

 www.trafford.com

North America & international
toll-free: 1 888 232 4444 (USA & Canada)
phone: 250 383 6864 ♦ fax: 812 355 4082

www.thecaterpillarsthatgrew.com

THE CATERPILLARS THAT GREW...
AND GREW...
AND GREW...

This book is dedicated to my wonderful husband Mike, who patiently waited for his dinner many an evening, while I wrote and illustrated this book.

No caterpillars or butterflies were harmed in creating this Adventure Book! Maddison and Mimi continue to learn as much as they can about the world of caterpillars and butterflies. They also seek information about the perfect plants to feed each type of caterpillar and attract many different species of butterflies. Butterflies make the world a more beautiful place for EVERYONE!

TABLE of CONTENTS

CHAPTER ONE 5
Maddison's Amazing Caterpillar Adventure
- Butterfly House in Navarre 14
- Water Park in Navarre 18

CHAPTER TWO 30
Butterfly ... Flutter By

CHAPTER THREE 37
The Amazing Life Cycle of a Butterfly!
- METAMORPHOSIS 42
- THE GREAT ESCAPE 47

For more information please visit our website at:
www.thecaterpillarsthatgrew.com
May This Metamorphous Never End!

The early morning sun was just peaking over the fence. Suddenly, a beautiful black butterfly fluttered over the fence and into Mimi's flower garden. It caught Maddison's eyes as it took a sip of nectar from a bright pink zinnia blossom.

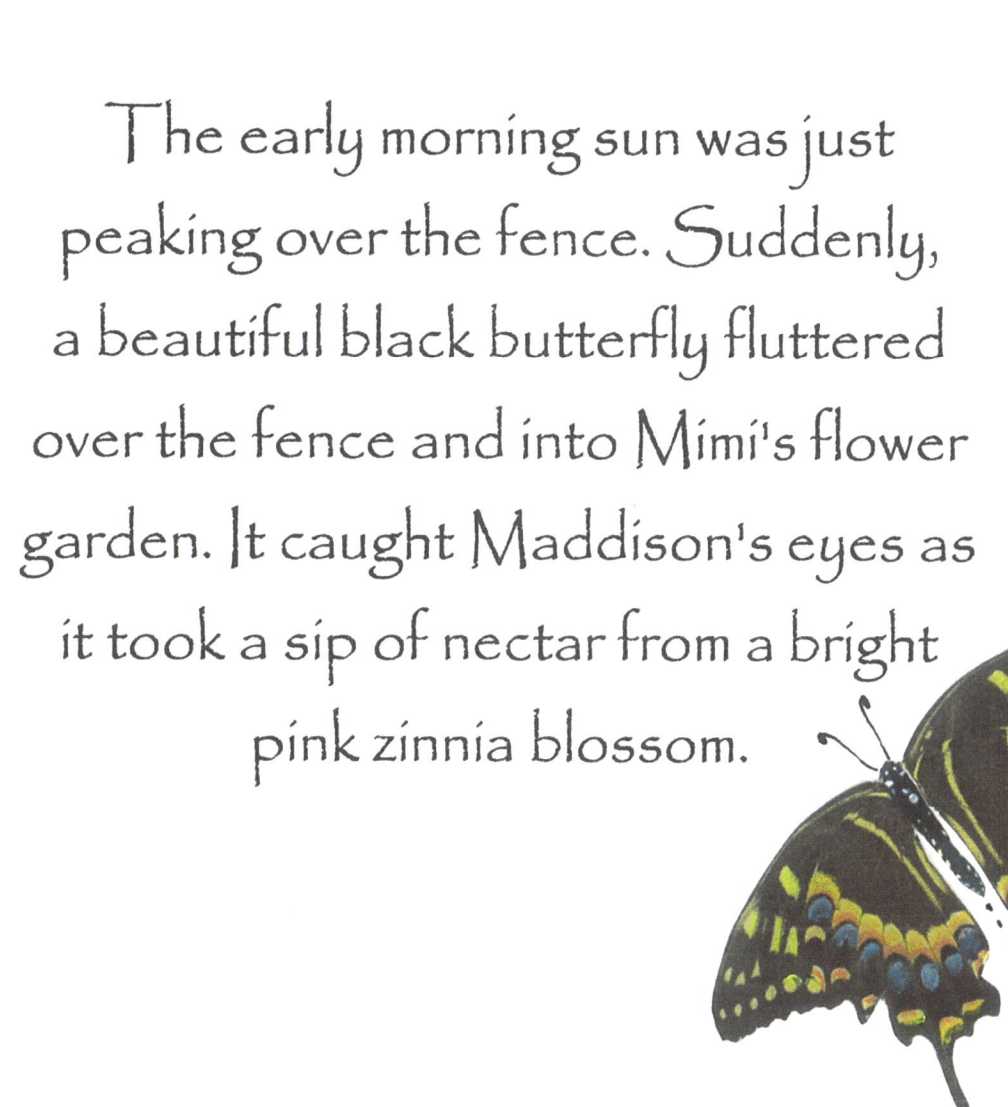

Then quick as a wink it flew
up . . .
up . . .
up . . .
over the fence and away!

It was only a few days later, as Maddison was helping her grandmother Mimi to water the herb garden, when they noticed many tiny caterpillars

MUNCHING…
yes, really MUNCHING…

on a parsley plant with curly leaves! Before long, they noticed three more tiny black and white caterpillars

MUNCHING…
this time on a lacy dill plant!

Maddison loved to help Mimi in the garden! Often she would help to water the plants. And finding these cute little caterpillars were even more fun!

Maddison used her magnifying glass that Papa had given her to get a close-up view of the caterpillars. She giggled as she watched the caterpillars MUNCH on the dill! Maddison pinched a dill leaf. It smelled just like DILL PICKLES! She wondered if caterpillars would like dill pickles as much as she did!

Maddison checked on her caterpillar friends every day. But all those caterpillars did was
MUNCH… MUNCH… MUNCH!
And of course, they just grew
FATTER … and FATTER… and FATTER!
Mimi told Maddison that there were many creatures that liked to EAT caterpillars, such as hungry birds and wasps and green garden lizards. So Mimi made a special butterfly house by covering a square frame with netting. Now Maddison could keep some caterpillars safe from hungry predators, looking for a tasty caterpillar snack!

Mimi and Maddison decided to share the rest of the caterpillars with other children. Maddison loved to visit the Navarre Beach BUTTERFLY HOUSE, next to a great PLAYGROUND, the WATERPARK and the BEACH! There were beautiful flowers, butterflies, and even ducks in a pond! There were swings and monkey bars filled with laughing KIDS…KIDS…KIDS! Maddison loved playing with KIDS! So they took a pot of parsley to the BUTTERFLY HOUSE. All the ENTOMOLOGISTS were very excited and thanked Maddison and Mimi for so many fat Black Swallowtail Caterpillars!

An Entomologist named Marge showed Maddison how to pick up a caterpillar and gently stroke his back to make him stick out two orange horns on the top of his head. Its horns are used to frighten away hungry predators. These caterpillars can also emit a stinky odor that many insects and animals don't like one bit! Because it is only a tiny bit of odor, people don't really notice it much. After placing the caterpillar back on the parsley plant, Marge carried the pot to the nursery where the caterpillars were safe to MUNCH and GROW!

Maddison and Mimi waved goodbye and skipped off to the Waterpark for a cool splash with lots of other KIDS! And Maddison's mother, Karen, took off work early to join them for an afternoon of cool refreshing FUN!

Maddison continued to check on her three caterpillar friends each day. One morning she looked very puzzled as she told Mimi that there were only two fat caterpillars munching on her dill plant. Mimi looked all around the butterfly house, then pointed way up inside. The third caterpillar was just sitting very still with a tiny thread holding him in place. Mimi told Maddison that the caterpillar was ready to form a CHRYSALIS. Once the caterpillars have eaten enough, they spin and wrap themselves into a paper-like sleeping bag. They won't eat again until they are butterflies, hungry to to sip nectar from the garden flowers.

The next day, all the caterpillars had changed quite a bit! Since yesterday, the remaining two fat caterpillars had joined the first and crawled away from the dill plant. Maddison used her Magnifying Glass to see how much they had changed! Each caterpillar was now a pale tannish-brown CHRYSALIS, just leaning back against a tiny silk thread. Mimi told Maddison that it would take about two weeks before any more changes would occur. So they took out the dill plant and placed the butterfly house in the warm barn, away from any bad weather.

It was just two weeks later when Mimi called out in excitement for Maddison to come quickly...

"Hurry!" exclaimed Mimi. "I want you to see something amazing!"

"What is it?" asked Maddison as she came skip-hopping into the barn.

"Come see for yourself," replied Mimi as she slowly opened up the netting to the butterfly house.

To Maddison's amazement, a lovely BLACK SWALLOWTAIL BUTTERFLY was hanging from tissue paper-thin wrappings, all that was left of its CHRYSALIS!

"Oh... how beautiful it is!" exclaimed Maddison.

Soon, the butterfly began to flap its wings as it tried to fly. Mimi knew that the butterfly might damage its fragile wings if it kept trying to fly while inside the netting. So she told Maddison to gently reach inside the box and offer her finger to her butterfly friend. It only took a minute before the butterfly stepped onto Maddison's finger, then began to walk up her arm. Maddison giggled as the butterfly suddenly flitted over to her bright pink shirt! "Does he think I am his Mama?" asked Maddison. "I don't think so," replied Mimi. "Perhaps your shirt looks like a big pink flower filled with tasty nectar!"

Maddison and Mimi sat down quietly next to the flower garden. Slowly, the butterfly opened its wings to catch the warm sunbeams. After awhile, the butterfly began to flap its wings as if to wave goodbye to its new people friends! Then off it fluttered!   UP…
and UP…
and UP it flew, stretching its beautiful black and yellow and cornflower blue wings gracefully as it flew in spirals and swirls across the sky! Then quick as a wink, the butterfly landed on a pink cosmos, thirsty for a tasty sip of sweet nectar!

Maddison pretended that she was a beautiful butterfly, flapping her arms gracefully as she danced around the yard. Maddison loved to dance. And ballet was Maddison's favorite form of dance. She imagined fluttering from flower to flower just like a butterfly. Maddison's aunt, Julianne, was a very talented ballerina who practiced hours each day so that she could one day become a famous Prima Ballerina! Maddison took ballet lessons from Julianne every Friday, hoping she could also become a Ballerina! Mimi smiled as Princess Maddison danced away in a Beautiful Butterfly Ballet! What a wonderful way to spend the day!

CHAPTER TWO

Butterfly . . .
      Flutter by!
Just look at all the Butterflies!
Gracefully waving arms like wings,
Maddison flutters as she sings!
So many beautiful butterflies,
    Sipping nectar . . .
      Then fluttering by . . .
  Soaring UP . . . UP . . . UP . . .
      into the sky!
Butterfly . . .
      Flutter by!

Mimi and Maddison knew that they would have to keep a close eye on the butterfly house. So they moved it to the porch so that perhaps they could catch one of the other butterflies as it came out of its CHRYSALIS. Sure enough, when Maddison checked the box the next morning, she could hardly believe her eyes. A tiny black head was just popping through the brown skin of the CHRYSALIS! Then out came skinny little legs and a tangle of wet wings! Soon, the wet butterfly was just hanging from its CHRYSALIS, with its wings pointing downward. Every once and awhile, it would flap its wings a little bit. Otherwise, it was very still.

Mimi explained what was happening. "Once the butterfly comes out of its CHRYSALIS, it must allow time for its wings to dry. By hanging from the empty CHRYSALIS, it can shift fluid from its body to the framework of its wings. As this framework of veins fills, the body thins out so that the weight of the butterfly is balanced. As the veins in the wings fill with fluid, the wings begin to stiffen as they dry. As the wings dry, the butterfly begins to flap them gently. It is then that one begins to see the unique colors and markings. But until the wings are dry, we can't disturb the butterfly. The wings must properly fill before drying, or the butterfly won't be able to fly."

RING...RING...RING...
Mimi got up to answer the telephone. When she came back, she shared some good news with Maddison. Marge called from the Navarre Beach BUTTERFLY HOUSE to say that there were Black Swallowtail Butterflies coming out of each CHRYSALIS every day!

"Would you like to go see the butterflies?" asked Mimi. "Yes, please," said Maddison! "I would like that... can we go right now?" "Let's get dressed and pack a lunch," said Mimi! "YIPEE," said Maddison! They hurried to dress and get in the car. Off they drove, LICKETY-SPLIT!

When they got to the Navarre Beach BUTTERFLY HOUSE, there were Black Swallowtail Butterflies everywhere! Of course, there were many other kinds of butterflies fluttering around, too! Amongst a zillion flower blossoms were hundreds of hungry butterflies searching for a tasty bit of nectar! "That butterfly kissed me!" said Maddison as a Black Swallowtail Butterfly fluttered by her face. "It tickles!" said Maddison. Mimi snapped a few pictures of Maddison as she posed amid a flurry of butterflies. Maddison took Mimi's hand and they both smiled . . . what a lovely way to spend the day . . . watching so many butterflies dancing in the sunshine!

CHAPTER THREE

Butterfly... Flutter by...
You fly so high up in the sky,
Gossamer wings to flutter and play
Sipping nectar throughout the day!

Once a common caterpillar,
Just MUNCHING all day...
Now you are sleepy
So you follow God's way!

Time for change... Yawn and Sigh
It's off to sleep, but who knows why...
God forms your wings to fly away,
So you can wake up as a butterfly
One beautiful day!

"Where do caterpillars come from?" asked Maddison. "Well," said Mimi, "A butterfly lays an egg, then out hatches a caterpillar." "Yes, but where do butterflies come from?" Maddison persisted. "Hmm," replied Mimi, "the butterfly comes from a fat caterpillar that made a chrysalis." "But where do we get caterpillars?" insisted Maddison. "OK . . . I think that I see where all this is really going," Mimi replied thoughtfully. "And I think the best way to answer your questions is to go way back in time, back to when God created the World." "As a child, this was the way things were explained to me."

"Father GOD created everything! He loved to create beautiful new things. So one day, He came up with a special plan, because everything that God makes is here for a special reason." "Like me?" asked Maddison. "Oh yes, exactly like you," replied Mimi with a smile. "Father God created the Universe as a great big canvas. He added many galaxies with planets and moons, then he splashed everything with twinkling stars. God picked Earth to decorate, creating mountain peaks and crystal clear streams, aqua blue oceans and sandy white beaches, magnificent gardens filled with every sort of animal and plant one could possibly imagine!"

"Then Father God decided to create the first man and the first woman. God wanted someone to talk with and share all His amazing creations. He created man in His own image, and gave man a loving heart just like God's own. Then he created woman so that man would have a Soul Mate. God gave woman a loving heart, too. And He named the first couple Adam and Eve, and gave them beautiful Garden Earth to live in! All the plants and insects and animals lived in harmony. I can just see God smiling as he decorated Garden Earth with happy songbirds, and thousands of colorful butterflies sipping sweet nectar as they flittered about the garden!"

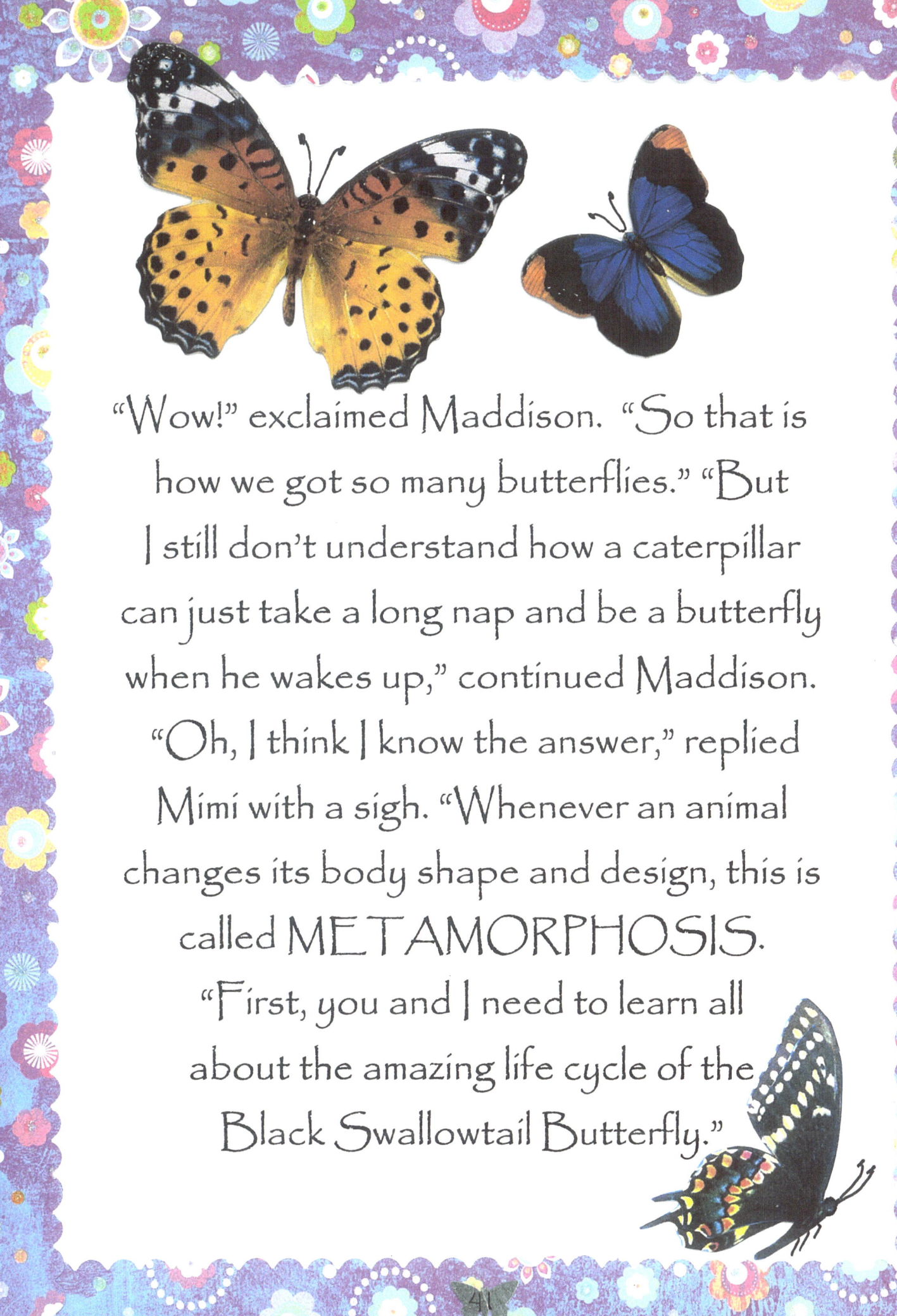

"Wow!" exclaimed Maddison. "So that is how we got so many butterflies." "But I still don't understand how a caterpillar can just take a long nap and be a butterfly when he wakes up," continued Maddison. "Oh, I think I know the answer," replied Mimi with a sigh. "Whenever an animal changes its body shape and design, this is called METAMORPHOSIS.

"First, you and I need to learn all about the amazing life cycle of the Black Swallowtail Butterfly."

METAMORPHOSIS

"The female Black Swallowtail Butterfly looks everywhere for the right type of host plant to lay its eggs on. Black Swallowtail caterpillars only eat certain types of leaves, such as dill, parsley, fennel and plants coming from the carrot family. So this is where the female must lay her eggs. In about one week, a tiny baby caterpillar will eat his way out of his egg case, which is full of nourishment. Today is HATCH DAY!"

"Happy Hatch Day to You!"

"For about 20 days, the caterpillar has only one job… MUNCH! and MUNCH… and MUNCH… you get the picture."

"Oh yes, he does get to take a few naps!"
"GOD gave the caterpillar an incredible sense of time... sort of a built in ALARM CLOCK!"
"I tell you this because the caterpillar will have to shed his skin about 5 times before he can spin his CHRYSALIS and take a nap. He has to do that before he can become a butterfly. He can't take any shortcuts or skip any steps. He has to be right ON TIME and do every step or he can never be anything more than ant food! And ants LOVE to eat tasty caterpillars that don't follow GOD's plan!"

"Back to MUNCHING...
because a caterpillar will MUNCH thousands of times his own weight, he will just get FATTER and FATTER! And the FATTER he gets, the tighter His [Cat Suit] will get. As people grow, they have to get new clothes in a bigger size. A caterpillar will shed his very tight skin every time he grows to the next stage, called INSTAR. The caterpillar will go through five INSTAR stages before his built-in clock tells him to spin a CHRYSALIS."

"When it is time for a new Cat Suit, the caterpillar will swallow air, stretching his skin until it literally splits apart!"

"This is called MOLTING."

"Finally, one day the caterpillar has grown to be the absolute biggest CAT (my nickname for caterpillar) that he can be! Then (YAWN) somehow his built-in clock tells him that it is time to stop MUNCHING, spin a CHRYSALIS, and just HANG OUT for awhile!"

"Here is the part where GOD really does some beautiful work! An amazing change is about to take place. What once was a simple worm just doing its job eating some plants to keep them from growing too big and taking over the garden, will become a totally different creature! And you thought he was just taking a nap in his sleeping bag!"

"This is when Metamorphous takes place. As we look at a CHRYSALIS, we see no movement, no head or legs, no mouth to eat with. There is little sign that anything inside is alive at all! But amazing things are happening inside the CHRYSALIS! Gradually the Cat body changes into a butterfly form, complete with antennae, butterfly eyes, and a slender body. Short stumpy legs are replaced by skinny new butterfly legs. Large delicate wings are formed. Instead of eating leaves like a caterpillar, the new butterfly will need a mouth-tube or PROBOSCIS to sip nectar from garden flowers."

THE GREAT ESCAPE

"One of the hardest days in the life of a butterfly might very well be the day it must escape from its CHRYSALIS!"

"The tiny butterfly body is tightly packed inside the CHRYSALIS, using every speck of space to form a perfect beautiful butterfly. It has taken 2 weeks to transform from a caterpillar to a butterfly, but now it is time to make an escape!"

"The CHRYSALIS splits, a tiny black head pops out, followed by six skinny legs. With much WIGGLING and WRIGGLING and SQUIRMING, a wet and crumpled looking butterfly takes pause to rest! A beautiful butterfly makes a Grand Debut into the world!"

"It is time to dry out. The butterfly must hang upside down, holding onto its empty CHRYSALIS case.
The butterfly works to pump air into its body, then body fluid into the veins of its still wet wings. Slowly, the wings stretch out and expand as the veins fill. About two hours later the wings are stiff and dry. Now the butterfly begins to test them out, flapping them slowly open and shut. Soon, the butterfly begins to flap its wings, becoming airborne in its first flight. Once the butterfly spreads its wings, it will instinctively fly UP...
and UP...
and UP...!

Maddison remembered that it was time to check the butterfly house on the back porch. There was still one more CHRYSALIS that had not split open yet. Sure enough, there were now two butterflies flapping their wings and looking for a way out of the box. It was easy to tell that each butterfly just wanted to spread its wings and soar through the air. Mimi reached in and gently cupped her hands over one of the butterflies. She showed Maddison how to open her hands and let the butterfly climb onto a stem of delicate flowers. Maddison did the same with the last butterfly! She was very careful to avoid hurting her beautiful butterfly as she set it next to the other.

Each butterfly extended its PROBOSCIS (feeding-tube) to sip the delicious nectar from some of the blossoms. Gently flapping its wings, each butterfly prepared for flight. As each playfully fluttered and flapped its wings, the butterflies soared upward rising HIGHER and HIGHER...
swirling UP...
and UP...
and UP to the sky!
Maddison and Mimi just grinned! And of course, Maddison flapped her wings and began to DANCE like a butterfly! WHAT A PERFECT WAY TO END THIS DAY!

PLEASE VISIT OUR WEBSITE AT:

www . thecaterpillarsthatgrew . com

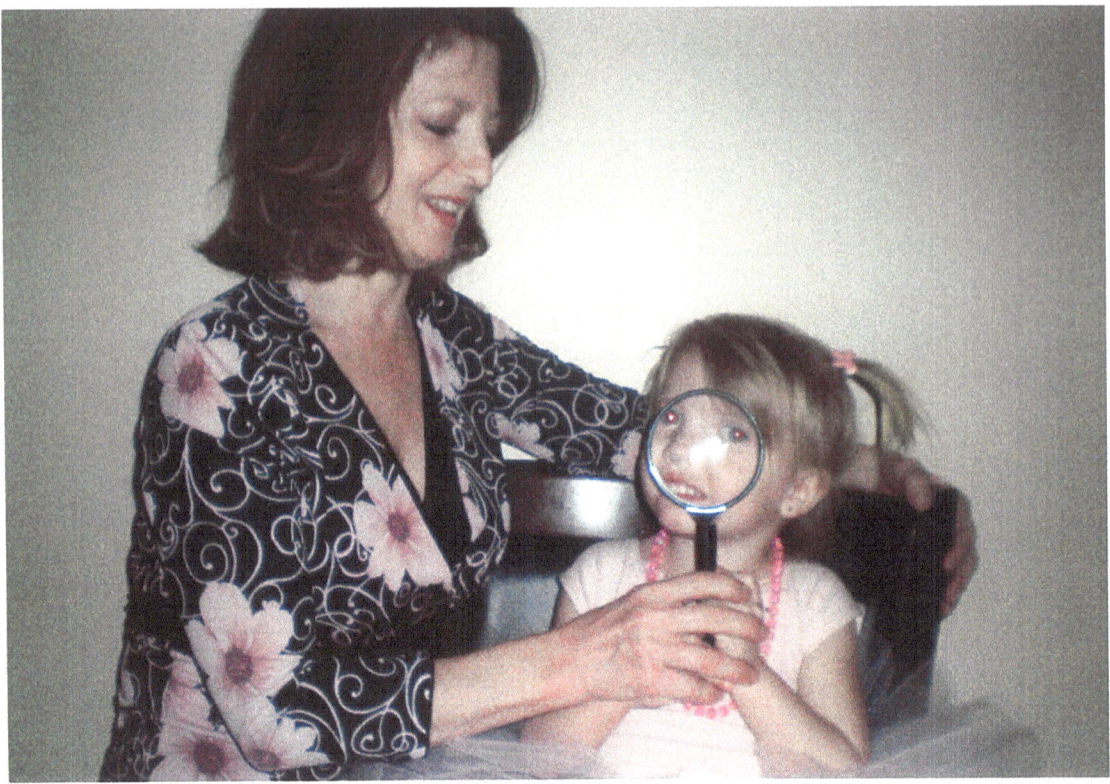

From pediatric nurse to mission volunteer in Honduras . . . to teaching children pottery, Robin Murray has spent much of her life caring for the needs of others. When chronic back pain forced her to slow the pace, quality family time became her focus.

Robin has always taken delight in expressing herself through a variety of art forms. She took delight in the ancient art of RAKU, a very labor-intensive method of firing pottery to achieve an array of striking pottery finishes. Now it is much easier to press and shape moist lumps of clay into much smaller pottery pieces . . . whimsical butterflies, green sea turtles and leapfrogs.

Instead of large beds filled with colorful garden plants, she shows her granddaughter Maddison how to dig and water pots of soft brown earth. Together, they grow pots of purple vincas, juicy red strawberries and rosy cherry tomatoes, ripe for the picking. More important are the pots of dill and parsley they plant, just waiting for the first Black Swallowtail Butterfly to flutter into the garden and lay her eggs.

These days, Robin chooses the descriptive power of words, accented with many bright colors, to perfectly illustrate an adventurous Children's Picture Book. And what better adventures to share than those from her own backyard, filled with MUNCHING caterpillars, fanciful butterflies, and the sweet sound of a grandchild's laughter! As Robin (Mimi) shares information about each life stage of the butterfly, Maddison peers through her magnifying glass to get a closer look. Through the use of her talents, Robin Murray seeks to encourage folks of all ages to promote and preserve much needed natural habitat for butterflies and their larva. Please join Robin and Maddison, and so many others, who strive to make the world a more beautiful place for all to enjoy!

MAY THIS METAMORPHOUS NEVER END!

www.ingramcontent.com/pod-product-compliance
Lightning Source LLC
Chambersburg PA
CBHW051055180526
45172CB00002B/640